TICKS

And What You Can Do About Them

Roger Drummond, Ph.D

Wilderness Press
Berkeley

FIRST EDITION May 1990
Second printing October 1990

Cover color photograph by Thomas Winnett
Cover design by Larry Van Dyke
Book design by Roslyn Bullas

Library of Congress Catalog Number 90-12236
International Standard Book Number 0-89997-116-4

Manufactured in the United States of America
Published by Wilderness Press
2440 Bancroft Way
Berkeley, CA 94704
(415) 843-8080
Write for free catalog

Library of Congress Cataloging-in-Publication Data

Drummond, Roger O., 1931-
Ticks and what you can do about them / by Roger Drumond.
p. cm.
ISBN 0-89997-116-4
1. Tick-borne diseases—United States. 2. Ticks—United States. 3. Ticks as carriers of diseases—United States. 4. Ticks—United States—Control. I. Title.
RA641.T5D78 1990
614.4'33—dc20 90-12236
CIP

Contents

Acknowledgements

The author wishes to acknowledge the help of a number of persons. Dr. Glen Needham, Institute of Acarology, The Ohio State University, Columbus, OH, provided the picture of tick mouthparts and the maps, and read the final draft. Dr. Jakie Hair, Department of Entomology, Oklahoma State University, Stillwater, OK, and Dr. James Oliver, Department of Biology, Georgia Southern University, Statesboro, GA, provided ticks for some of the pictures. Dr. Robert B. Craven, Center for Infectious Diseases, CDC, Ft. Collins, CO, provided information on Lyme disease. EcoHealth, Inc., Boston, MA, provided the plate of the actual size of the deer tick. Drs. Allen Miller, John George, Matthew Pound and Glen Garris and Mr. Stanley Ernst, Knipling-Bushland U.S. Livestock Insects Research Laboratory, Kerrville, TX, gave help and advice. Bill Miller, Dike Drummond, M.D., Cy Kendall, Jack and Rita Wilson, Maud Jennings, Dr. Sam Junkin, Ralph and Betty Pelton and Dr. Dan Sonenshine read drafts. Tom Winnett, Editor, Wilderness Press, provided constructive editorial advice. My most helpful advice, comment and criticism came from Ellen Drummond, friend, companion and wife.

Introduction

Have you or a member of your family found a tick on them?

Have you become aware of the diseases, especially Lyme disease and Rocky Mountain spotted fever, carried by ticks? Do you live in an area that has ticks? Are you planning to go to an area that has ticks? Have you read an article about ticks and the diseases they carry and want to know more? Are you looking for ways to keep ticks from getting on you and yours? This book tells about how and where ticks live, the diseases ticks carry and the problems they cause. It also gives detailed information on how you can protect yourself from ticks, how you can remove attached ticks safely and how you can use different materials and methods to control ticks.

Nature and all outdoors are to be enjoyed. As with any part of life, this enjoyment is not without its risks. One of these risks is the presence of ticks and tick-carried diseases, such as Lyme disease, Rocky Mountain spotted fever and several other diseases. A tick bite can cause severe irritation and even paralysis. The first 6 chapters of this book give you information on ticks and the diseases they carry. The last 4 chapters provide you with practical measures you can take to minimize the risks from ticks and tick-carried diseases. If you use these methods and materials you should be able to enjoy the outdoors with a minimum of risk of contacting ticks and suffering tick-carried diseases.

Chapter 1

Ticks And How They Live

Ticks are part of a large group of animals that have jointed legs and skeletons located on the outside of their bodies. This group, called arthropods, includes insects, crustaceans, myriapods and arachnids. Arachnids, in turn, include spiders, scorpions, mites and ticks. Ticks' only food is the blood they suck from their hosts (usually warm-blooded animals). There are about 800 kinds of ticks in the world. About 100 of these carry diseases of wildlife, livestock or people.

The front part of a tick consists of the "head" area and the mouthparts. The mouthparts have a central structure, the hypostome, which is shaped like a blunt harpoon, flat on the top and curved on the bottom, where many recurved barbs are located. A tick pushes its hypostome into a hole in the skin of a host that has been made by sharp teeth on the front of the hypostome. The barbs anchor the tick to the skin and make it difficult to pull the tick out. Plate 1 is a scanning electron photomicrograph of a tick's mouthparts. Some ticks also produce a cement-like substance that helps anchor them to the host. Sharp teeth at the front of the hypostome cut blood vessels under the skin, causing the blood to form a pool. The tick then sucks this blood into its gut through the hypostome. To keep the blood from clotting, ticks inject saliva containing a kind of anticoagulant into the blood pool. The saliva may contain disease organisms, too.

The rear part of a tick consists of the body, which is designed to expand as the tick takes in blood. When a tick feeds, the gut, inside the body, expands as it receives blood, and the accordion-like skin of the body unfolds and expands as the gut increases in size. Some ticks can increase their size 20-50 fold as they feed.

Ticks go through 4 stages in their life cycle—adult, egg, larva and nymph. The adult stage consists of sexually mature, 8-legged male and female ticks. The adult sexes usually differ in color, size and appearance—females usually being larger and more colorful than males. Females, and sometimes males, emit a chemical, called a pheromone, to attract the opposite sex. This pheromone allows adults to find each other. Adults find hosts, suck blood and mate. Some adults suck blood only once; others suck blood several times. Mating may take place on or off a host and before or after feeding.

Fully fed, mated females lay eggs. Some females lay all their eggs in a single mass, while others lay eggs in several small batches. Females die after laying all their eggs. Males live to mate several females.

Eggs hatch into tiny 6-legged larvae, often called seed ticks. A larva finds a host, attaches and sucks blood until full. When full, it detaches and drops to the ground. After a few days, the larva changes form, in a process called molting, and becomes a nymph. A nymph is larger than a larva, has 8 legs but still does not have sex organs. The nymph breaks out of the larval skin, finds a host, attaches and sucks blood. Some nymphs feed several times and molt to another nymph form before feeding and molting to an adult, while others feed only once and molt to an adult. The adult breaks out of the nymphal skin and finds a host.

Ticks have a unique system for finding hosts. They have sense organs on their front legs that detect carbon dioxide. Tick hosts, mostly warm-blooded animals, give off carbon dioxide. Using these sense organs, ticks crawl to the source of the carbon dioxide which is usually a host. Researchers and health officials use this behavior in order to collect ticks. They place a block of dry ice on a cloth, the dry ice sublimates to carbon dioxide and ticks crawl to the cloth where they are collected. Ticks may crawl 10-15 or more feet to a source of carbon dioxide. They do not fly to or jump on their hosts.

Ticks are divided into two families—"soft" ticks and "hard" ticks. Soft ticks do not have a hard shield on all or part of their body. The entire body of a soft tick can expand during feeding. Because of their shape and their wrinkled skin, they look like little raisins. Soft ticks usually live in sheltered places, such as burrows and nests of their hosts. Larvae feed only once, but feeding may last a week or more. Nymphs and adults usually feed several times, but feeding may last less than an hour. Mating takes place off the host after the adults have fed. Females lay several small batches of 50-200 eggs and then die. There are over 150 kinds of soft ticks.

Hard ticks have a hard shield that covers all or part of their body. In females, this shield covers only a small part of the front part of her back, and her body can expand greatly as she feeds. In males, this shield covers most of his back, and his body can expand only slightly as he feeds. Hard ticks are usually found in the open. Each stage feeds only once, for 3 days to several weeks. Mating usually takes place on the host before feeding. Females lay eggs in a single mass, which can be as large as 23,000 eggs, before they die. There are over 650 kinds of hard ticks.

Ticks are important because they carry diseases of wildlife, livestock and people. Although colonists in the eastern U. S. found a large number of ticks, there is no record that these settlers were affected by any tick-carried diseases. A traveler in New York state in 1749 wrote that it was impossible to sit down without being attacked by an army of ticks. Clearing of the eastern forests, which changed the environment, and a decrease in wild animals, which decreased the number of tick hosts, caused a decrease in the number of ticks. A visitor to the same area in 1872 noted that the common tick had become nearly extinct.

A different tick-human disease situation existed in the West. Indians in the Rocky Mountains recognized places with "evil spirits" (apparently ticks) that should be avoided or first traveled by squaws to remove the problem. Prospectors and settlers entered and lived in these areas and were fed upon by ticks that carried disease-causing organisms. The first recognized tick-carried disease of humans in the U.S. was Rocky Mountain spotted fever. Although this disease is still found in the Rocky Mountains, today more cases are reported in eastern than in western states.

More recently, a new disease of humans carried by ticks has been found in the U.S. This disease, called Lyme disease, is now a major health problem in some areas of the Northeast, the upper Midwest and the Pacific coast. Several other tick-carried diseases of humans are found in the U.S. The rickettsiae, viruses, bacteria and protozoa that cause these diseases develop inside ticks and are injected into persons while ticks feed.

Some ticks do not carry a disease but are a nuisance and can cause itching, swelling and redness when they "bite" a person. Actually, ticks don't bite; they attach to a host by sticking their mouthparts into the host's skin. If a tick is scraped off, sometimes the mouthparts break off and remain in the skin to cause irritation and infection.

Feeding by certain ticks can cause paralysis. This condition, called tick paralysis, is the result of a toxin injected into a person as the tick feeds. Paralysis spreads and can lead to death. Once the feeding tick is found and removed, the paralysis abates and the person usually recovers rapidly.

Chapter 2

Where Ticks Live

Ticks are found wherever their hosts are found. Like all parasites, ticks cannot live without hosts. Some ticks feed on only one kind of host, while others suck blood from many different animals. When not attached and feeding on their hosts, most hard ticks live on the ground in vegetation, such as grassy meadows, woods, brush, weeds, etc. They are found in the leaf litter and duff on the soil. In contrast, a unique hard tick, the brown dog tick, can live inside houses, kennels or wherever dogs are found. Most soft ticks live in the nests and burrows of their hosts and in houses, if their hosts live there.

Ticks are not uniformly distributed throughout the environment. Usually the largest numbers and kinds of ticks are found where there are the largest numbers and kinds of wild animal hosts. Most hosts are found around streams and lakes, meadows and other open areas in forests and also found in wooded areas in the prairies. Man changes the environment when he cuts down trees in the forest or puts fences and other barriers on the prairie. These changes can lead to more areas for wildlife and thus to a larger number of ticks. Often man puts domesticated animals on limited, fenced pasture areas. Some ticks in such areas that feed on wildlife may also feed on livestock. If so, their number may increase to many times the number supported by wildlife alone. In contrast, if man destroys the forest, plows all the prairie or overgrazes the land, ticks in the area may disappear because of the lack of hosts or of suitable places for ticks to live, molt and lay eggs.

There are many examples of changes in the situation with a tick-carried disease as a result of man's activities in the environment. One example is Rocky Mountain spotted fever. Rocky Mountain spotted fever became a problem with the settling of the West. Over time, man cleared much of the land and changed it to pastures and fields with limited wild areas. As a result, there was a decrease in the number and the distribution of the Rocky Mountain wood tick, which carries the disease, and a decrease in the number of cases of the disease in the Rocky Mountains. Now Rocky Mountain spotted fever in the West is found usually in rural areas where lumbermen, sportsmen, hikers, campers and others contact disease-carrying ticks.

In the 1940s, reports of cases of Rocky Mountain spotted fever shifted from the West to the Southeast. Since 1964, more than 90% of all cases of the disease have been recorded east of the Rockies. The reasons for this shift are that in the last 50 years people have been moving from cities to suburbs and rural areas, where they encountered disease-carrying ticks and that, with the abandoning of large farms and the regrowth of limited woods, there are more wildlife hosts and more places for ticks to live than in the past.

The tick that carries Rocky Mountain spotted fever in the East is the American dog tick. Suburbanites, in their wooded yards and small farms, created new areas for wildlife, good hosts for the American dog tick. They also brought their pets, especially dogs, which are excellent hosts for this tick. Pets can bring disease-carrying ticks into homes, so a person need not go outdoors to get Rocky Mountain spotted fever if a pet brings infected ticks into the house. Some people who had not even left New York City became infected with this disease. See Chapter 5 for more information on Rocky Mountain spotted fever.

Another example of a response of ticks to man's activities is the recent appearance of

Lyme disease in the U.S. Although known in Europe since the 1900s, this disease was first diagnosed in the U.S. in Old Lyme, Connecticut, in 1975 as Lyme arthritis. The organism that causes Lyme disease is found in the blood of many wild animals, especially deer and field mice. The disease is carried by 2 kinds of ticks, the deer tick, found in the Northeast and upper Midwest, and the western black-legged tick, found in the Pacific Coast states.

It is speculated that the deer tick was present from Minnesota to the Atlantic Ocean when this area was covered with trees and had a large number of white-tailed deer. During the 1800s, clearing the land reduced tick habitat and unregulated hunting reduced the number of deer, leading to a large decrease in the number and distribution of the deer tick.

But over the past 40 years, an increase in wooded areas, the regulation of hunting, and wildlife-conservation efforts caused an increase in tick habitat and in deer populations. As a result, there has been an increase in the number and the distribution of the deer tick. At the same time, people moved to the suburbs and to rural areas where deer, field mice and deer ticks are found. Deer ticks feed on mice and become infected with the disease organism. Infected ticks then feed on people as well as wildlife. As a result of increases in wildlife and infected ticks, exposure of more people to deer ticks, and greater public and medical awareness of the disease, there has been a dramatic increase in the number of reported cases of Lyme disease since 1975. See Chapter 4 for further information on Lyme disease.

Chapter 3

Important Ticks

In the U.S. 7 kinds of hard ticks and 5 kinds of soft ticks carry diseases, are a nuisance or cause paralysis. Table 1 lists the states, except Alaska and Hawaii, and ticks found there. Alaska has the Rocky Mountain wood tick outdoors and the brown dog tick indoors. Hawaii has the brown dog tick. The District of Columbia will have the ticks found in Maryland and Virginia.

Table 1. Ticks found in 48 States

	Hard ticks							Soft ticks	
State	Lone Star	Rocky Mt.	Pac. Coast	Amer. Dog	Deer	West. B-L.	Brown Dog	Relap Fever	Pajah-uello
AL	X			X			X		
AZ		X					X	X	
AR	X			X			X		
CA		X	X	X		X	X	X	X
CO		X					X	X	

	Hard ticks							Soft ticks	
	Lone Star	Rocky Mt.	Pac. Coast	Amer. Dog	Deer	West. B-L.	Brown Dog	Relap Fever	Pajahuello
State									
CT					X		X		
DE	X			X	X		X		
FL	X			X			X	X	
GA	X			X			X		
ID		X		X		X	X	X	
IL	X			X	X		X		
IN	X			X	X		X		
IA	X			X			X		
KS	X			X			X	X	
KY	X			X			X		
LA	X			X			X		
ME					X		X		
MD	X			X	X		X		
MA					X		X		
MI				X	X		X		
MN				X	X		X		
MS	X			X			X		
MO	X			X			X		
MT		X		X			X	X	
NE		X		X			X		
NV		X	X			X	X	X	
NH					X		X		
NJ	X			X	X		X		
NM		X					X	X	
NY				X	X		X		
NC	X			X			X		
ND		X		X			X		
OH	X			X			X		
OK	X			X			X	X	
OR		X	X	X		X	X	X	
PA				X	X		X		
RI					X		X		
SC	X			X			X		
SD		X		X			X		
TN	X			X			X		
TX	X			X			X	X	
UT		X				X	X	X	
VT					X		X		
VA	X			X	X		X		
WA		X		X		X	X	X	
WV	X			X	X		X		
WI				X	X		X		
WY		X					X	X	

Lone Star Tick

Amblyomma americanum

The unfed female of this hard tick has a white mark, "the lone star," on the shield in the middle of her back (Plate 2). The rest of her body is reddish-brown. The smaller male is the same color, with pale lacy white markings on the rear edge of his back (Plate 2). The much smaller larvae and nymphs are reddish-brown.

Adults are active in spring and early summer. Larvae are active in late summer. Nymphs are active from spring until fall. Adults and nymphs survive winter buried in the soil. Larvae and nymphs feed on many small animals; adults and nymphs feed on larger animals. All stages are found on deer and also readily feed on man.

The lone star tick is found throughout the southeastern third of the U.S. (Map 1). Many are found in the Ozarks. In eastern Oklahoma, it is estimated that this tick kills 20-40% of white-tailed deer fawns. Feeding causes the skin where ticks attach to swell and bleed. Fawns with many ticks on them go blind, die or become so weak they are killed by predators.

A person who walks through vegetation that contains a clump of lone star larvae may find many tiny ticks crawling on him. They are usually found attached to the skin under

Map 1

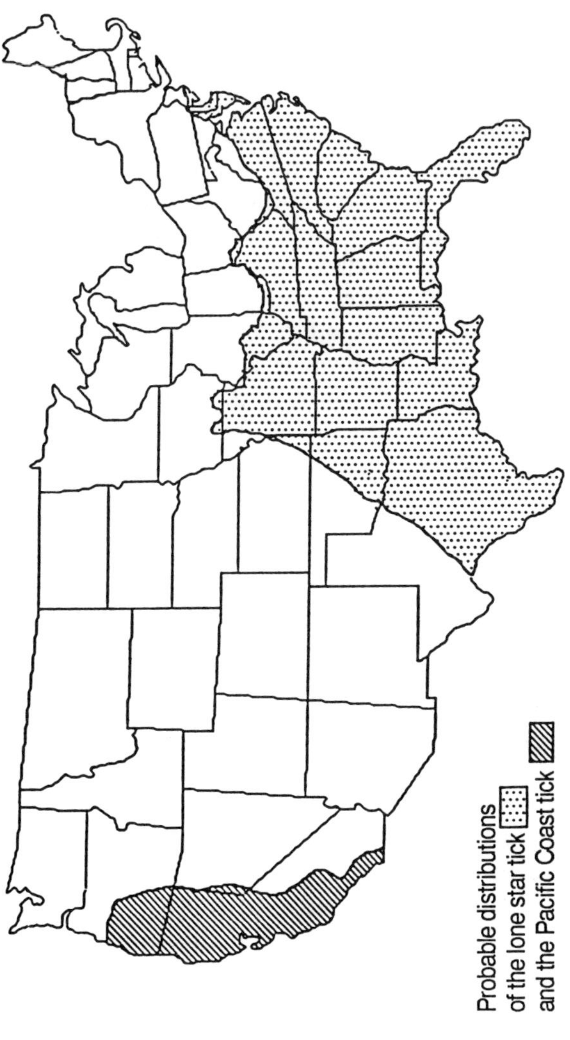
Probable distributions
of the lone star tick
and the Pacific Coast tick

tight places, such as bra straps and belts. Ticks crawl on the skin until they are stopped by tight clothing, and then they attach at that place. Larvae are usually scraped off during bathing or showering, if that is done soon after contact with the ticks. If not removed, larvae can cause a rash and itching. Adults and nymphs also are found attached in tight places. The attachment site may become inflamed and infected.

Lone star ticks give tularemia to people. Because they are often found on people, lone star ticks have been looked at as carriers of other diseases. True, organisms that cause Rocky Mountain spotted fever and Lyme disease have been found in lone star ticks, but there is no proof this tick gives these diseases to people.

Rocky Mountain Wood Tick
Dermacentor andersoni

This hard tick looks like the Pacific Coast tick and the American dog tick (Plate 3). The unfed female has silvery-grey markings on the shield on the front third of her back; the rest is reddish-brown. The unfed male has silvery-grey markings on the shield on all of his back. Larvae and nymphs are dark reddish-brown.

Adults usually spend winter buried in the soil and become active in early summer. Larvae also are active in summer. Nymphs may be active in summer or they may spend winter in the soil and become active in spring. Larvae and nymphs feed on a variety of small animals. Adults feed on many large animals.

The Rocky Mountain wood tick is found in the Rocky Mountain states west to the Pacific Ocean and south to New Mexico and Arizona (Map 2). Along the Pacific coast, it is usually found only in the higher elevations. It is the prime carrier of Rocky Mountain spotted fever in the West and also carries tularemia and Colorado tick fever. It is the most important cause of tick paralysis in the U.S.

Pacific Coast Tick

Dermacentor occidentalis

This hard tick looks like the Rocky Mountain wood tick and the American dog tick (Plate 3). The unfed female has a silvery-grey pattern on the shield on her back; the rest of her body is reddish-brown. The unfed male has silvery-grey markings on all his back. Larvae and nymphs are dark brown.

Adults are active in late spring. Larvae and nymphs are active in late spring and early summer. Larvae and nymphs feed on a variety of small animals. Adults feed on many large animals.

The Pacific Coast tick is found in California, southeastern Oregon and northwestern Nevada (Map 1). It carries tularemia and is a suspected carrier of Rocky Mountain spotted fever and Colorado tick fever.

Map 2

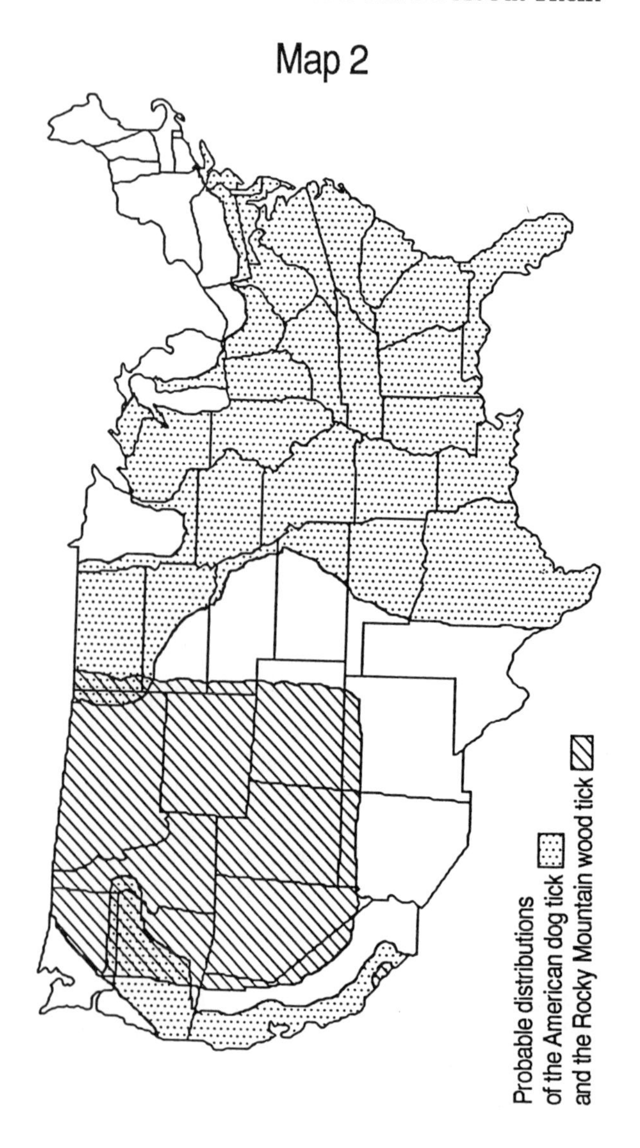
Probable distributions
of the American dog tick
and the Rocky Mountain wood tick

American Dog Tick
Dermacentor variabilis

This hard tick looks like the Rocky Mountain wood tick and the Pacific Coast tick. In many places this tick is called the wood tick. The unfed female (Plate 3) has silvery-grey markings on the shield on her back; the rest of her body is reddish-brown. The unfed male (Plate 3) also has silvery-grey markings on all his back. Larvae are a dull yellow and nymphs are reddish-brown.

Adults overwinter in the soil and are active from spring to fall. Larvae are active in spring and summer. Nymphs are active in summer and may overwinter. Larvae and nymphs feed on mice and many other small animals. Adults prefer dogs but feed on many large animals, including man.

The American dog tick is widely distributed in the eastern half of the U.S. and is found on the west coast (Map 2). It is the most important carrier of Rocky Mountain spotted fever in the U.S. It also carries tularemia and causes tick paralysis.

Deer Tick
Ixodes dammini

This hard tick was described in 1979. In the Northeast, this tick is called the deer tick because it was first found on deer; in the Midwest, it is called the bear tick because it was

Deer Ticks *(Ixodes dammini)*
Actual Size

larvae nymph adult male adult female

first found on bears. It looks like the western black-legged tick. The unfed female (Plate 4) is dark reddish-brown and has no markings on her back although the shield is visible. The unfed male (Plate 4) is dark reddish-brown. Males and females both have black legs. Larvae are very dark colored and nymphs are dark reddish-brown. The deer tick is a very small tick. Larvae are about the size of the period at the end of this sentence. Nymphs are about the size of a poppy seed, and adults are about 1/8 inch long. The actual size of all stages of the deer tick is shown in Figure 1.

Adults may feed in fall or in warm weather in winter and early spring. Both adults and larvae may winter in the soil and find hosts in spring. Nymphs are active all summer. Larvae prefer to feed on mice. Nymphs feed on many animals. Adults feed on deer and other large animals. All stages, especially nymphs and adults, have been found feeding on people.

The deer tick is found in the Northeast from Maryland to Maine and in the northern Midwest (Map 3). Each year it is discovered in more and more locations. This tick is the most important carrier of Lyme disease in the U. S. and is the only known carrier of human babesiosis.

Western Black-legged Tick
Ixodes pacificus

Unfed females and males look like deer ticks (Plate 4). Females are dark reddish-brown and males are dark brownish-black. Both have black legs. Larvae and nymphs are dark reddish-brown. This tick is about the same size as the deer tick (Figure 1).

Adults are active from November through May. Larvae and nymphs are active in late spring to summer. Larvae and nymphs feed on lizards and other animals; adults feed on many hosts.

This tick is found along the Pacific coast as well as in parts of Idaho, Nevada and Utah (Map 3). It is the carrier of Lyme disease in the western U.S.

Brown Dog Tick
Rhipicephalus sanguineus

The unfed female of this hard tick is brown (Plate 5). The shield on her back is darker than the rest of her body. The unfed male is a uniform dark brown (Plate 5). As females

Map 3

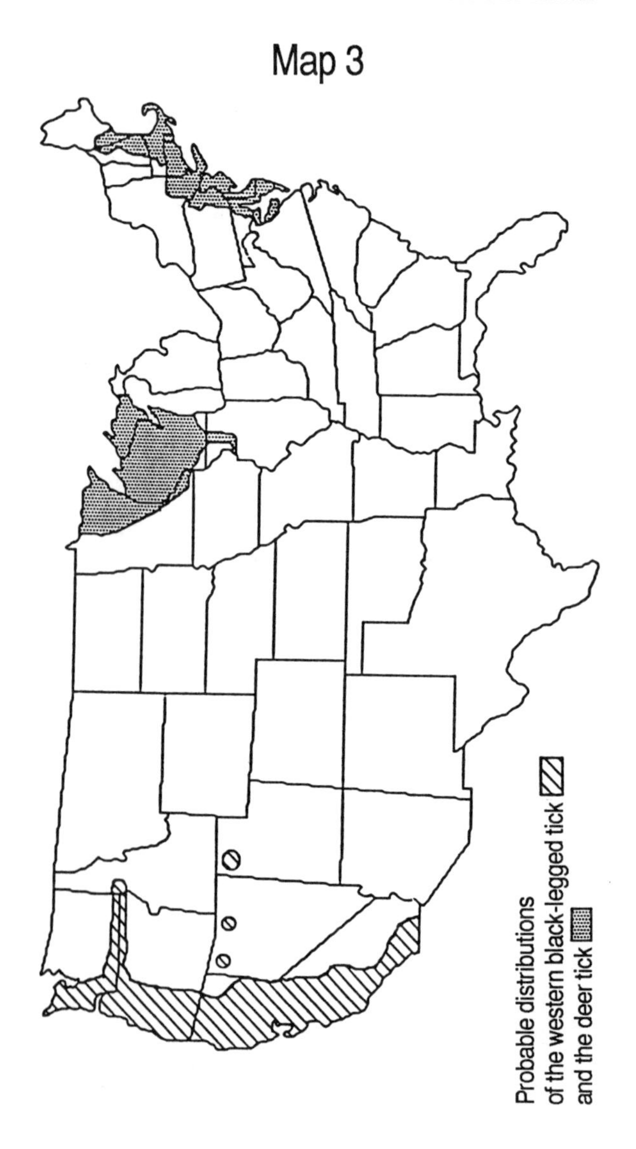
Probable distributions
of the western black-legged tick
and the deer tick

feed, they expand as they take in blood. As their skin stretches they change to a dark grey color. Unfed larvae and nymphs are dark brown.

The brown dog tick is found throughout the U.S. wherever you find dogs. All stages of this tick can be found indoors. Full females lay eggs in secluded places, and full larvae and nymphs drop off dogs and molt in any sheltered place. Unfed larvae, nymphs and adults crawl on and may try to attach to people, although attachment is uncommon. Although suspected of carrying human ehrlichiosis, brown dog ticks are not known to carry any diseases of people.

Relapsing Fever Ticks

Ornithodoros hermsi, O. parkeri, O. talaje, and *O. turicata*

These four kinds of soft ticks carry relapsing fever; all look alike and have similar life cycles. Adults are grey to pale blue, and their oval-shaped soft bodies have a rough texture (Plate 6). Larvae and nymphs are grey.

Relapsing fever ticks are found widely scattered west of the Mississippi River, in nests and burrows of many animals. *O. turicata* is also found in Florida in burrows of wild and feral pigs and in nests of the gopher tortoise.

Relapsing fever ticks can live without feeding for 5-10 years. They attack man only when they are hungry and man invades their habitat. The bite of these ticks can be painful.

Pajahuello Tick

Ornithodoros coriaceus

This soft tick looks like relapsing fever ticks (Plate 6). It is found in bedding grounds of deer and cattle. All stages feed on deer, cattle and other hosts, including man.

The pajahuello tick is found only in California and was originally called "pajaronela" by natives of the Santa Lucia mountain range in coastal California. It carries no human diseases, but is a problem because of the severe reactions to its bite. There is usually a localized reaction of pain and swelling where the tick attaches. These symptoms usually disappear in 2 days, but often there is a small knot at the site, which usually disappears in 1-2 weeks. Certain persons allergic to the bite may have reaction with swelling, pain and redness that may require medical attention and take several weeks to disappear.

Chapter 4

Lyme Disease

Lyme disease was diagnosed in the U.S. in 1975 in Old Lyme, Connecticut, where an unusual arthritis was named Lyme arthritis. In Europe, this disease, found since 1910, is called *erythema chronicum migrans,* which means chronic migrating red rash. Lyme disease was found in Wisconsin in 1970 and in California in 1978. It is caused by a spirochete named *Borrelia burgdorferi,* a kind of bacterium. In the U.S. the number of reported cases of Lyme disease has increased almost every year since 1982. Lyme disease is now the most common tick-carried disease in the U.S., but it is concentrated in certain parts of the country. In 1988 over 4500 cases were reported. Over half of these cases were reported from New York State, and about 80%

of the cases in New York State were reported from Westchester and Suffolk counties. Table 2 shows the reported cases of Lyme disease from 1982-1988. It is speculated that many more cases were not reported.

Table 2. Reported cases of Lyme disease, 1982-88, from 4 states with highest numbers, rest of the states and all the states. Preliminary data furnished by the Center for Infectious Diseases, Ft. Collins, CO.

State	1982	1983	1984	1985	1986	1987	1988	Total
NY	170	267	466	235	482	877	2637	6134
CT	135	73	83	699	0	215	362	1967
NJ	57	70	155	175	219	257	500	1433
WI	58	69	176	135	162	358	246	1204
Rest of the states	71	116	236	504	523	664	830	2943
All states	491	595	1536	2748	1836	2371	4575	13681

Distribution

Lyme disease is concentrated in 3 areas of the U.S. Most cases are found from Maryland to Maine. The second area is Wisconsin and Minnesota and the third is California and Oregon. Over 94% of all reported cases have been found in nine states—California, Connecticut, Massachusetts, Minnesota, New Jersey, New York, Pennsylvania, Rhode Island and Wisconsin. Lyme disease is found in 34 other states. Only seven states—Alaska, Arizona, Hawaii, Montana, Nebraska, New Mexico and Wyoming had not reported cases through 1988. Information on cases in 1989 is not yet available.

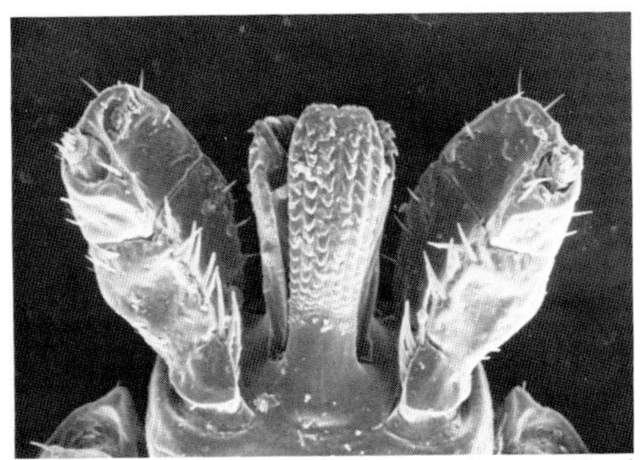

Plate 1. A scanning electron microphotograph of the mouthparts of the American dog tick. The central structure is the hypostome.

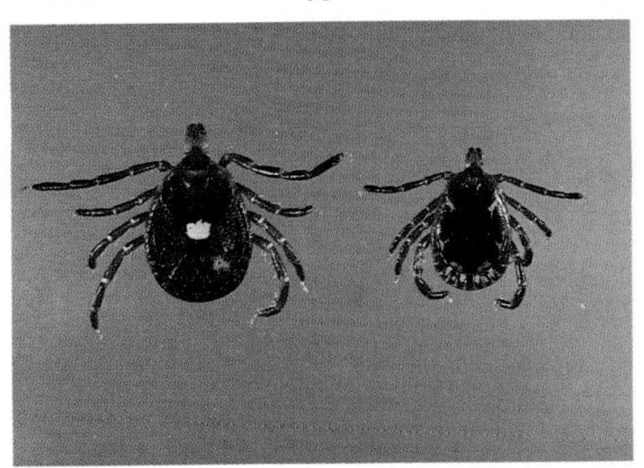

Plate 2. Color photograph of adults of the lone star tick. Female left and male right. Actual size about 1/4 inch long.

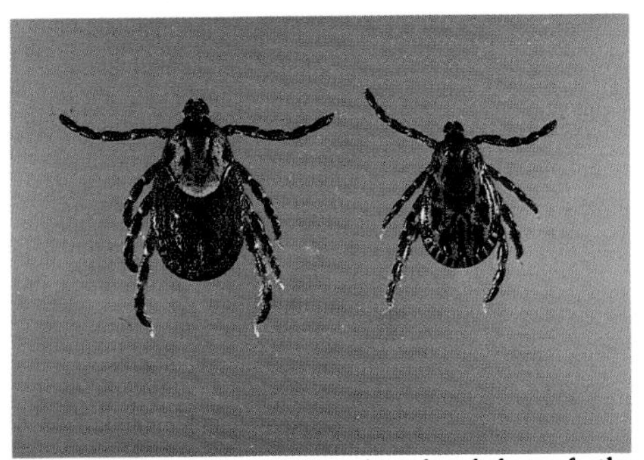

Plate 3. Color photograph of adults of the American dog tick. Female left and male right. Actual size about 1/4 inch long.

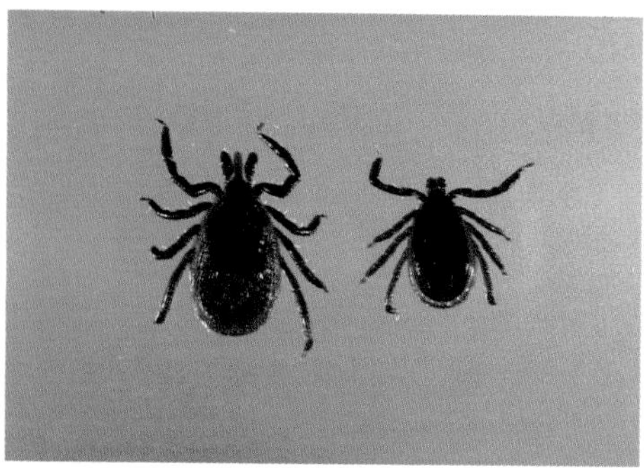

Plate 4. Color photograph of adults of the deer tick. Female left and male right. Actual size about 1/8th inch long.

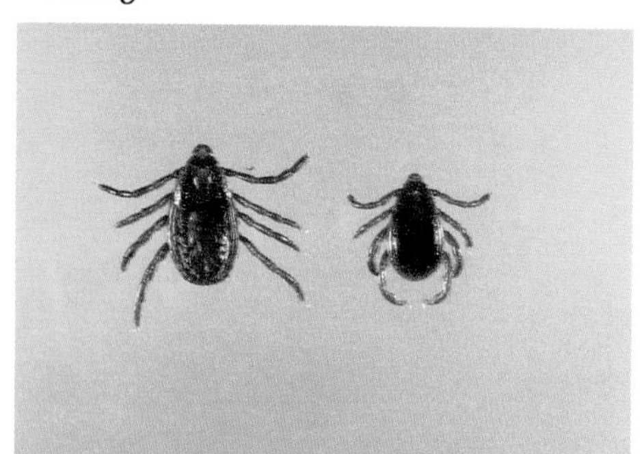

Plate 5. Color photograph of adults of the brown dog tick. Female left and male right. Actual size about 3/16th inch long.

Plate 6. Color photograph of an adult relapsing fever tick. Actual size about 1/4 inch long.

Symptoms

Lyme disease is called the great imitator because its symptoms imitate the symptoms of so many other diseases. The symptoms of Lyme disease appear in 3 stages. First symptoms usually appear from 2 days to a few weeks after a person is bitten by an infected tick. Symptoms usually consist of a ring-like red rash on the skin where the tick attached. Unfortunately, this rash appears in only 60-80% of infected persons. Often the rash is red around the edges with a clear center. The person also has flu-like symptoms of fever, fatigue, chills, headaches, a stiff neck and muscle aches and pains. Rashes may be found some distance away from the original rash. These symptoms often disappear after several weeks.

Second-stage symptoms, which occur weeks or months later, include meningitis, severe headaches, encephalitis, paralysis of facial muscles, abnormal heart beat, numbness, withdrawal, loss of confidence, lethargy and other symptoms. These symptoms may last for several weeks or months.

Third-stage symptoms, which occur months or years later, include arthritis, especially in large joints, with swelling, pain and stiffness. Arthritis may become chronic, and in children may be mistaken for juvenile rheumatoid arthritis. Other symptoms include fatigue, numbness and loss of memory. Third-stage symptoms may mimic symptoms of multiple sclerosis and Alzheimer's disease.

Early diagnosis of Lyme disease is based on finding a tick and the rash, or if there is no rash, the severe flu-like symptoms. A blood test also can usually tell if a person is infected. Antibiotics given during the first stage can eliminate or decrease the severity of the second- and third-stage symptoms. Antibiotics can also treat second- and third-stage symptoms. Persons in areas with Lyme disease must be alert to ticks on them and to a rash or severe flu-like symptoms in the summer. Prompt medical attention can prevent or lessen the long-term effects of Lyme disease. Lyme disease is not considered to be fatal, but the arthritis and other third-stage symptoms may be very severe and long-lasting.

Tick Carriers And The Disease Cycle

The deer tick, in the Northeast and the Midwest, and the western black-legged tick, in the Rocky Mountains and along the Pacific coast, carry Lyme disease. These are small ticks (see Figure 1, page 20) and hard to see and detect. Larvae and nymphs take in spirochetes as they feed on infected mice. Infected larvae molt to infected nymphs and infected nymphs molt to infected adults. Infected ticks feed on people, who get Lyme disease. More cases of Lyme disease in people are the result of the feeding of infected nymphs than of infected adults. Most cases of Lyme disease occur between late spring and early fall when people are outdoors and ticks are most active.

Lyme disease is such a new disease that every year we learn more about it. In laboratory tests, the lone star tick and the American dog tick did not transmit Lyme disease after feeding on infected hosts and are not considered carriers. In contrast, the black-legged tick, *Ixodes scapularis*, became infected and transmitted spirochetes to other hosts. Black-legged ticks have also been found infected with Lyme disease spirochetes in nature, but as yet there is no evidence that this tick, found scattered throughout the Southeast and Southwest from Maryland to Texas, carries Lyme disease to people.

Lyme disease has been diagnosed in pets, especially dogs, and in horses and dairy cattle.

The presence of Lyme disease in places that do not have known carrier ticks is a mystery. It is important that the symptoms of Lyme disease be recognized by all persons and be well known to doctors in areas that are not now known to have Lyme disease, so that the disease can be detected and treated early. There still is much to learn about diagnosis, treatment and carriers of Lyme disease.

Chapter 5

Rocky Mountain Spotted Fever

Rocky Mountain spotted fever is the oldest known tick-carried disease of people in the U.S. It was first recognized as a disease in the 1890s. Because this disease was found in the Rocky Mountains and infected persons had a rash with red-purple-black spots, it was called Rocky Mountain spotted fever. This disease is caused by a rickettsia, *Rickettsia rickettsii*. The number of reported cases of Rocky Mountain spotted fever increased almost every year until 1948, when antibiotics caused a sharp decline in reported cases. Fewer than 200 cases were reported in 1959. The number of reported cases increased slightly during the 1960s and rapidly during the 1970s. The number of cases peaked at 1192 in 1981 and 619 were reported in 1989.

Distribution

Until 1930 this disease was reported only in the Rocky Mountains although it may have occurred undetected in the eastern U.S. Then cases were reported from the southeastern Atlantic coast. By the late 1930s more cases were reported in eastern, southeastern and southwestern states than in Rocky Mountain states. In 1963, more than 90% of all cases were recorded east of the Rocky Mountains. In 1989, more that 20 cases were reported from Arkansas, Georgia, Missouri, North Carolina, Oklahoma, South Carolina and Tennessee. Rocky Mountain spotted fever is found in all states except Alaska, Hawaii and Maine. More persons are now exposed to infected ticks in the South and East than in Rocky Mountain states.

In the past in the Rocky Mountains, the disease was found mostly in men who spent their time in the woods, exposed to infected ticks. Now, there are fewer ticks, areas with ticks are more limited and, because the disease is found in remote areas, people, mostly men, are not so exposed as in the past. But now in the South and East, the disease is found among children and women who contact infected ticks in their yards and recreation areas and from pets. In 1987 four cases of Rocky Mountain spotted fever were reported from the Bronx. The victims had not traveled outside New York City within the 3 weeks before they showed symptoms of the disease.

Symptoms

The first symptoms of Rocky Mountain spotted fever include severe headache, chills, fever and general aches and pains. In a few days, in most cases, a rash of reddish-purple-black spots appears on the soles of the feet, ankles, palms of the hands, wrists and forearms. This rash may later spread to the trunk, neck and face. At the end of a week, without treatment, a person may become highly agitated, develop insomnia, become delirious or go into a coma. Recovery in less severe cases may take weeks or months. Antibiotics given early in the disease are very effective. With treatment, recovery takes 2-4 days and there are usually no lasting effects. Without treatment, there can be lasting effects and 3-5% of infected persons die.

Even though this disease has several well defined clinical symptoms, it is difficult to determine if an illness is Rocky Mountain spotted fever. Clinical diagnosis is confirmed by blood tests, but treatment should not wait for these tests.

Tick Carriers And The Disease Cycle

The Rocky Mountain wood tick and the American dog tick carry Rocky Mountain spotted fever. As a larva, nymph or adult feeds on an infected host, it can become infected. An infected larva molts to an infected nymph, and an infected nymph molts to an infected adult. Nymphs and adults feed on people, who become ill with Rocky Mountain spotted fever. Also an infected female tick can pass rickettsiae to her eggs. Since she can lay thousands of eggs, many infected larvae can be produced by one infected female.

The rickettsia that causes Rocky Mountain spotted fever can also get into a person's body through the skin—a tick bite is not necessary. People can get Rocky Mountain spotted fever if they crush a tick and get blood or fluids from the tick on their skin or in a cut or sore. People can also get the disease if they handle an infected wild animal or pet and get urine, blood or other fluids from the animal on them.

In the Rocky Mountains, Rocky Mountain spotted fever is a disease of late spring and early summer, when adults of the Rocky Mountain wood tick are active. In contrast, Rocky Mountain spotted fever east of the Rockies is found throughout spring and summer, when adults of the American dog tick are active.

Chapter 6

Other Diseases

Tularemia

Tularemia, also called rabbit fever, first described in the western U.S. in 1911, is caused by a bacterium, *Francisella tularensis.* Reported cases of tularemia declined from a high of 2291 in 1939 to less than 200 per year in the 1970s. Since then, the number of reported cases has risen slightly. There were 144 cases reported in 1989. Tularemia is found in almost every state. In 1989, more than 10 cases were reported from Arkansas, Missouri and Oklahoma.

Symptoms of tularemia include chills and fever, loss of appetite, weakness, swollen lymph nodes and an ulcer-like wound at the site of the tick bite. Symptoms become more severe if the disease is not diagnosed and treated with antibiotics. The disease can be diagnosed by identifying the bacterium and by specific blood tests. There are a few deaths each year.

The Rocky Mountain wood tick, the American dog tick and the lone star tick carry tularemia. Ticks feed on infected mice, rabbits and other animals, and infected ticks give bacteria to hosts as they feed. Infected females can pass bacteria to their larvae through their eggs. Other blood-sucking insects, such as horse flies, deer flies and mosquitoes, also carry the disease.

People can also get tularemia when they contact diseased animals. Rabbit hunters need to be very careful with rabbits they kill. Tularemia organisms can enter through the skin if a person crushes an infected tick. Most cases of tularemia are found in spring and summer when ticks and other carriers are active, and in fall, when hunters contact infected rabbits.

Colorado Tick Fever

This disease, first thought to be a variety of Rocky Mountain spotted fever, was shown to be a separate disease in the 1940s. Colorado tick fever is caused by a virus. Since 1946, the number of reported cases of Colorado tick fever has ranged from 200 to 400 per year. Many cases are not reported because of the mild symptoms. It has been reported in

California, Colorado, Idaho, Montana, Nevada, New Mexico, Oregon, South Dakota, Utah, Washington and Wyoming. The disease is found from about 4,000 to 10,000 feet altitude.

Symptoms include a sudden high fever, fatigue, chills, severe headache and muscle aches. The fever may last for 1-3 days and return suddenly in 1-2 days. There is no purple-red-black rash, as seen in Rocky Mountain spotted fever, although sometimes there is a faint rash. Other symptoms may include stiff neck, irritation, loss of memory and loss of coordination. No specific drug is used to treat Colorado tick fever, although treatment for headache and muscle pains may be given. Victims usually recover within 2 weeks without treatment, but in some cases recovery may take several months. This disease can be diagnosed by virus identification and blood tests.

Only the Rocky Mountain wood tick carries Colorado tick fever. Most cases occur in spring and early summer, when nymphs and adults of this tick are active.

Human Babesiosis

The first case of human babesiosis in the U. S. was found on Nantucket Island in 1969. This disease is caused by a malaria-like protozoan, *Babesia microti*. By 1980 a total of over 100 cases had been reported. Only about 10-15 cases are reported each year. Cases are found only in Massachusetts, New York, Rhode Island and Wisconsin.

Symptoms include fatigue and loss of appetite followed by fever and chills, headache and muscle aches. In severe cases there is anemia, jaundice and blood in the urine. Symptoms usually go away without treatment, but may be most severe in older persons and those without a spleen. Drugs used to treat malaria can be used to treat babesiosis. Babesiosis can cause death in older patients. The disease can be diagnosed by finding disease organisms in the blood or by specific blood tests.

Only the deer tick carries human babesiosis. Most cases are found from May through July when people are outdoors and deer ticks are active.

Relapsing Fever

Relapsing fever, found only in the western U.S., is caused by a *Borrelia* spirochete carried by the 4 kinds of relapsing fever ticks. Only a few cases are reported each year. Relapsing fever is usually found in limited outbreaks in which a few persons in a single location all get the disease at the same time.

Relapsing fever takes its name from the fever that comes and goes in cycles. High fever lasts from 2 to 9 days and is followed by 2 to 4 days without fever; then the person relapses into another period of high fever. Periods of fever may show up months after the start of the disease. Antibiotics can be used to treat relapsing fever. There are no records of death as the result of this disease. Relapsing fever can be diagnosed by finding disease organisms in the blood and by blood tests.

Relapsing fever ticks are found in the nests of mice, squirrels, chipmunks and other small animals. These animals often build their nests in cabins or shelters. When the animals are killed or driven out, the remaining ticks become hungry and feed on people who happen to be there.

Q Fever

Q fever, also called nine-mile fever, is short for query fever. This disease, found widely scattered in the world and rarely in the U.S., is caused by a rickettsia, *Coxiella burnetii*. Usually the disease appears in distinct outbreaks among persons in the same area or occupation.

Symptoms include severe and persistent headaches and a low pulse rate. Symptoms may suddenly occur and last for 1-4 weeks. There is no rash and the disease is usually mild. Antibiotics can be used to treat the disease. Diagnosis is made by the identificaton of disease organisms in the blood.

Q fever organisms have been recovered from the Rocky Mountain wood tick, the Pacific Coast tick and the lone star tick.

Laboratory studies have shown that the Rocky Mountain wood tick can become infected by feeding on infected hosts and infected females can pass the organism onto their larvae through the eggs. The Q fever organism is very infective and can get into the body through the skin. Persons can even get Q fever by inhaling disease organisms. Outbreaks of Q fever have occurred among laboratory workers, veterinarians and farmers in the U.S.

Human Ehrlichiosis

The first case of human ehrlichiosis in the U.S. was found in Arkansas in 1986. This disease is caused by a rickettsia, *Ehrlichia*. The most common *Ehrlichia* organism in the U.S. is *Ehrlichia canis*, the cause of canine ehrlichiosis. The specific organism in the U.S. that causes the disease in man has not been identified. Cases of this disease have been reported from Alabama, Arkansas, Georgia, Mississippi, Missouri, New Jersey, North Carolina, Oklahoma, South Carolina, Tennessee, Texas and Virginia.

Symptoms can range from very mild to so severe that the person needs hospitalization. Symptoms include fever, headaches, aches and pains in the joints and muscles, chills, loss of appetite, eye pains, nausea and vomiting. There is usually no rash. Antibiotics are used to treat human ehrlichiosis. Diagnosis can be done only by specific blood tests.

The tick that carries this disease in the U.S. has not been determined. The brown dog tick is suspected as a carrier, but this tick seldom feeds on people. Other suspected carriers are the lone star tick, American dog tick and deer tick.

There have been very few cases of human ehrlichiosis in the U.S. Information on the disease and its carriers is scarce; more should be available in the future on human ehrlichiosis.

Tick Paralysis

Tick paralysis is not a disease but a tick-caused "problem" first reported in the Rocky Mountains in 1912. It is not caused by a disease organism but by a toxin a tick injects as it feeds. Cases of tick paralysis have been found in most Rocky Mountain and Pacific coast states, especially Montana, Idaho, Oregon and Washington. Scattered cases are found in Virginia, the Carolinas, Georgia, Kentucky and Tennessee.

Symptoms usually appear about 4-6 days after a person is exposed to ticks. Paralysis usually starts in the hands and feet. Then there is loss of coordination and activity in the legs and arms. Next there is paralysis of the face, with slurred speech and uncontrolled movements of the eyes. Finally there is irregular breathing. The inability to breathe leads to death. These symptoms appear over a period of 8 days and are most severe in children.

Paralysis is always associated with feeding by a tick attached on any part of the body, although the head is most common. Once the tick is removed, symptoms disappear in reverse order of appearance. Usually there are no lasting aftereffects of tick paralysis. In cases of unknown paralysis, a thorough search of the body should be made to look for a feeding tick.

The Rocky Mountain wood tick is the most important cause of tick paralysis in the U.S. The American dog tick also causes tick paralysis. Most cases in the West occur in April-June, when adults of the Rocky Mountain wood tick are active. In the South and East, most cases are found in March-August, when adults of the American dog tick are active.

A summary of locations of important ticks and diseases or problems caused by these ticks is presented in Table 3.

Table 3. Summary of ticks, locations and diseases or problems

Tick	Location	Disease or problem
Lone star	Southern 1/3 of the U. S.	Nuisance Tularemia Q fever? Human ehrlichiosis?
Rocky Mountain wood	Western mountains	Rocky Mountain spotted fever Tularemia Colorado tick fever Tick paralysis Q fever?
Pacific Coast	West Coast	Tularemia Rocky Mountain spotted fever? Colorado tick fever?
American dog	Eastern 1/2 of the U.S. and the West Coast	Rocky Mountain spotted fever Tularemia Tick paralysis
Deer	Northeast and upper Midwest	Lyme disease Human babesiosis
Western black-legged	West Coast	Lyme disease
Brown dog	Everywhere	Human ehrlichiosis?
Relapsing fever	Scattered in the West and Florida	Relapsing fever
Pajahuello	California	Painful bite

Chapter 7

Personal Protection

You can keep from getting tick-carried diseases and tick-caused problems by making sure you do not get ticks on you. You could stay away from woods, recreation areas, lawns and other places ticks live. However, ticks can even be found in the house, brought there by pets that go outside and bring them in. The brown dog tick can live in a house or any other place dogs are found. Although this tick does not carry any major human disease, they can be a nuisance. If brown dog ticks are hungry and cannot find a dog to feed on, they may crawl onto and try to attach to people.

You usually find hard ticks outdoors. Their normal hosts are mice, rabbits, birds, deer and other wild animals. They get on people when they find a person instead of their normal host. Some ticks find a host by awaiting on grass or shrubs along roadsides or paths used by animals. When a host passes, the ticks get on it. Other ticks find a host by crawling to the host when it is resting or slowly passing by. It is difficult to avoid ticks if you are there when they are seeking a host.

You rarely find the soft ticks that carry relapsing fever outside the nests and burrows of their normal hosts. They leave these places only if they are hungry and their normal hosts are gone. Usually people have to be very close by before these ticks find and attack them. Relapsing fever outbreaks have occurred when people have stayed in cabins and shelters that also housed small animals with relapsing fever ticks. You will usually find the pajahuello tick in resting places and bedding grounds of cattle and deer under trees and shrubs. People contact this tick when they walk through or rest in these areas.

You will have to make some decisions about what you are going to do about ticks and tick-carried diseases. If you think the risk of getting a tick-carried disease is great, you may decide to stay away from an area and not expose yourself to the danger. If you decide to expose yourself to ticks and diseases, you can take none, some or all of the measures described in this book to reduce the exposure. The measures will do you no good if they are not used.

Proper Clothing

You can prevent or reduce the numbers of ticks that attach to you by seeing them and removing them before they attach. You can wear light colored clothing so that ticks are visible on your clothes. Wear a long-sleeved shirt that fits tightly at the wrists and neck, and tuck the shirt into your pants. Wear long pants and tuck the pant legs into your boots or your socks. You could even use masking tape to tape the pant legs tightly to your socks, shoes or boots. You do all of this to keep ticks on the outside of your clothes, so you can see them and remove them before they reach your skin and attach. Long-sleeved shirts and long pants can be very hot in the summer, but their use is justified if there is a threat of a tick-carried disease where you are and you are serious about keeping ticks off you.

You should check your clothes while outdoors, and especially before entering your house to make sure you do not carry ticks inside. Once inside, it is a good idea to place all of your clothing on a sheet and examine it carefully for ticks. If possible, a better idea is to wash the clothes as soon as possible. Washing removes the ticks from the clothes. Ticks in your house may attach later, when you are not expecting it.

You should inspect your pets when they come in, to make sure they do not have ticks on them. If you find ticks on your pet, you can remove them carefully (Chapter 8) or you can treat your pet with approved insecticides (Chapter 9).

Despite all your efforts, ticks may still get to your skin. Some people can feel ticks crawling on them; most cannot. A good practice, as soon as possible after coming indoors, is to shower or bathe and to check your body for ticks, especially in areas that have hair or where clothing was tight. You should inspect yourself visually or by hand. If possible have another person examine you to make sure you have not missed any ticks. You will usually see bigger ticks, both nymphs and adults, and pick them off if they have not attached, or else carefully remove them (Chapter 8) if they have already attached. It is difficult to see and remove larvae ("seed ticks"). Usually you notice them only after they have attached and their feeding has caused the attachment site to itch. You can scrape them off or you can treat the places they are attached with medications used to treat the body for lice.

It is especially important to protect children with proper clothes. Thoroughly bathe and inspect children, and remove any attached ticks promptly.

Repellents Applied to Clothes

You can decrease the number of ticks that attach by treating your body or clothing with a repellent. The most common, EPA-approved and effective repellent for ticks, chiggers and other pests is "deet" (*N,N*-diethyl-*meta*-toluamide). You can buy this repellent under names like Off! and Cutters throughout the U.S., and it is available in sticks, bottles and spray cans. Thoroughly read the label of any repellent product to make sure it contains deet, and apply as directed for maximum effectiveness and safety.

You can apply deet directly to your skin to repel ticks, mosquitoes and other biting insects. Ticks that crawl on the treated area are irritated by the repellent and drop off. However, deet is most effective against ticks when applied to clothing from a spray can. Hang up your shirt and pants or lay them on the ground, hold the can about a foot or less from the clothes and spray each side of each piece for about 15 seconds. You must treat both sides thoroughly. If you are already dressed, you can spray your front pretty well, but you should have someone else spray your back. You must avoid getting the product in your eyes. Treat footwear, socks and pants cuffs for protection at these critical places.

In tests, volunteers applied a repellent that contained 20% deet to their shirts and pants and walked through or sat in tick-infested areas for 1/2 - 1 hour. They determined protection by recording the numbers of ticks that reached their skin. The treatment provided about 85% protection against the lone star tick, 94% protection against the American dog tick and 85% protection against the deer tick. In other tests, a repellent that contained 30% deet afforded 93% protection against the deer tick. In all these tests, the volunteers taped their treated pants to or tucked them inside their socks or boots. Ticks had to stay on the outside of the clothing, where they were exposed to the repellent.

There have been a few reports of adverse toxic reactions to deet, such a seizures, allergic responses and skin irritation. To minimize the possibility of these reactions you should: 1) apply deet sparingly to your skin, 2) avoid applying high-concentration products (those with greater than 30% deet) to the skin, particularily of children, 3) not inhale or ingest products or get them into the eyes, 4) apply repellent to clothes to reduce exposure to deet, 5) avoid treating parts of children's hands that may contact eyes or mouth, 6) do not treat wounds or irritated skin, 7) use repellent sparingly and 8) wash skin after coming indoors. If a reaction occurs, wash the skin and take the repellent can with you when you see a physician. Do not apply repellents that contain deet to rayon, spandex or synthetic fibers other than nylon. They may damage furniture finishes, plastics, painted surfaces and plastic watch crystals.

Toxicants Applied to Clothes

Only one toxicant is approved for you to apply to your clothes to control ticks. This product, Permanone™ Tick Repellent, contains permethrin, an insecticide. You can purchase Permanone Tick Repellent only in an aerosol spray can. The product is called a repellent because it kills ticks so rapidly that the ticks appear to be repelled.

Permanone Tick Repellent is approved for sale in Alabama, Arkansas, Colorado, Connecticut, Delaware, Florida, Georgia, Iowa, Kentucky, Louisiana, Maryland, Massachusetts, Michigan, Minnesota, Mississippi, Missouri, Montana, New Jersey, New Mexico, North Carolina, North Dakota, Ohio, Oklahoma, Pennsylvania, South Carolina, Tennessee, Texas, Virginia, West Virginia and Wisconsin. Approval is pending in Arizona, Illinois, Kansas, New York, Oregon and Wyoming.

Label instructions call for thorough treatment of clothing before dressing. You must allow the clothing to dry before putting it on.

Volunteers applied Permanone Tick Repellent to their clothing and tested the product against ticks as they did in the tests with deet. The treatment provided 100% protection against the lone star tick, the American dog tick and the deer tick. In other tests, Permanone Tick Repellent gave 100% protection against the Pacific Coast tick and the western black-legged tick. In laboratory tests, Permanone Tick Repellent applied to a cloth killed the pajahuello tick.

Permanone Tick Repellent usually kills ticks. Ticks that have crawled on treated clothing for only a few seconds drop off and most of them die. Usually a single treatment will protect against ticks for a day or longer. It is best to treat clothing daily or at least every other day for maximum effectiveness.

You must not apply Permanone Tick Repellent directly to your skin, face or eyes. If you accidentally should get some on you, you must wash it off with soap and water. If you get it into your eyes, flush them with plenty of water. Carefully read the label on the Permanone Tick Repellent can for other precautionary statements.

Chapter 8

How to Remove Ticks

Despite all your efforts, you have found a tick attached to you or yours. What do you do? You should remove a tick as soon as possible after you discover it. Ticks in the process of attaching are easier to remove than firmly attached ticks. More importantly, it may take several days for an infected tick to give a disease to its host, so the sooner you find and remove an attached tick, the less chance you have of getting the disease the tick is carrying.

We all have heard of special methods to remove attached ticks. Some claim these "folk" methods will cause ticks to "back out" or detach. There are two reasons why you must remove attached ticks properly. The first is that you must be certain to remove all the mouthparts from your skin. A tick attaches by inserting its hypostome, shaped like a blunt harpoon with many recurved barbs (Figure 1), into the skin. If you remove the tick improperly, the hypostome may break off, remain in the skin and become a source of irritation and infection. The second reason is that the organisms that cause Rocky Mountain spotted fever, tularemia and Q fever can enter the body through the skin, and if you crush or break an attached tick, the disease organism may get into your body. You must remove attached ticks so that you do not crush or puncture them and do not break off their mouthparts.

The lone star tick has the longest mouthparts of all the ticks in this book and is the most difficult to remove. The degree of difficulty in removing a tick depends upon the length of the mouthparts, the amount of cement secreted, the stage of the tick (mouthparts of adults are larger than those of nymphs, which are larger than those of larvae) and the length of time the tick has been feeding. Some persons react to an attached tick by swelling at the site, and the swelling may make it more difficult to remove the tick.

Researchers allowed adult American dog ticks to attach to sheep. After 3-4 days, they covered some ticks with petroleum jelly, some with clear nail polish and others with rubbing alcohol. They treated a few by lighting a wooden kitchen match, blowing it out and touching the tick with the hot, smoking end of the match. They found that none of these folklore methods made the ticks detach. They removed other ticks by grasping them with protected fingers or with blunt curved forceps or tweezers as closely as possible to where the mouthparts entered the skin, then steadily pulling the ticks from the skin. They found no crushed ticks or broken mouthparts. They obtained the same results with ticks that had been attached for only 12-15 hours.

Researchers also allowed adult lone star ticks, which have much longer mouthparts than American dog ticks, to attach to sheep. They grasped the ticks with blunt curved forceps or tweezers as close to the skin as possible. Some ticks they pulled steadily away from the skin. Others they pulled with a jerk, and yet others with a twisting motion. Finally they pulled some parallel to the skin. They removed all the ticks successfully—none was crushed and mouthparts were not broken.

Since this research, a new device, The Tick Solution, is available in the U.S. It is a specially designed spring-loaded, pointed forceps. You place the forceps around the tick as close to the skin as possible and release the spring, which drives a cylinder that closes the forceps firmly around the tick. Then rotate the tick several times, until it comes free of the skin.

The best way to remove an attached tick is as follows:

1. If possible, use blunt curved forceps, tweezers or The Tick Solution. If you use your fingers, cover the tick with waxed paper, plastic wrap, tissue or paper toweling.
2. Grasp the tick as close to the skin as possible.
3. With forceps or tweezers, remove the tick with a steady pull away from the skin—do not jerk or twist the tick. If using The Tick Solution, rotate the tick several times in both directions to loosen it and then pull it away from the skin.
4. Take great care not to crush or puncture the body of the tick or to get any fluids from the tick on you.
5. Kill the removed tick by placing it in alcohol or flush it down the toilet. If you are in an area with Lyme disease, you may wish to save the tick in a bottle with moist paper to give to a physician for examination.
6. After you have removed the tick, disinfect your skin with alcohol and wash your hands with soap and water.

Chapter 9

Area Control With Insecticides

Other than personal protection, the most common way to fight ticks is to treat the places ticks live with insecticides. You can apply insecticides safely and effectively to kill ticks.

The Environmental Protection Agency (EPA) has approved a number of insecticides you can apply to houses, kennels, yards and other places to control the brown dog tick. Most of these insecticides are approved to control the brown dog tick only, while a few of these insecticides are also approved to control the brown dog tick and other ticks outside the house. A very few other insecticides can be used only by pest control operators, public-health officials or other responsible personnel.

Your local county agricultural extension agent or advisor, public health personnel or other officials should have a list of insecticides you can use to control ticks. They should also have information on how to use approved insecticides safely and effectively. Because of changes that take place in the list of approved insecticides, this book presents only general information on effective and safe use of insecticides.

Insecticides are sold in pet shops, feed stores, garden shops, nurseries, grocery stores, discount stores and many other places. In most of these stores, you will find many products but, usually, no one who knows how to select and use them. It is up to you to choose the correct one.

The label on an insecticide can, bottle or package contains all the information you need in order to select and use it safely and effectively. Read the label thoroughly so you will know if the product is approved to control the brown dog tick only, or to control it and other ticks. The label will tell how and where to apply the insecticide. If you don't follow the instructions exactly, the treatment may be ineffective, damage the environment or be toxic to you, pets and wildlife. The label also contains information on hazards to humans and animals, practical treatments for accidental exposure, precautions and environmental hazards, and storage and disposal of the container.

You will find some insecticides sold as ready-to-use products you can apply directly from the container. Others are liquid concentrates or wettable powders you must dilute with water before use. The label gives exact instructions on how to dilute the concentrate or powder.

To control the brown dog tick, you must treat the dog as well as where the dog lives. Treating the dog will kill attached ticks and treating where the dog lives will kill ticks that are laying eggs, molting or looking for a host. You can treat a dog by washing or dipping it, or you can use an insecticide collar. You must thoroughly spray baseboards, window and door frames, cracks and floors. You should remove all bedding and replace it with fresh bedding after treating the dog and where the dog lives and runs. If dogs run outdoors, treat the ground and other places they live and play. You may have to treat several times before you get rid of brown dog ticks.

You can control the other hard ticks listed in Chapter 3 by spraying insecticides to kill ticks looking for hosts. Treat weeds, shrubs, underbrush and nearby grassy areas. Treat with enough volume to wet plants and the ground thoroughly. Spray along footpaths and roadsides, where ticks congregate. Do not permit children or pets to go into treated areas until the spray has dried. Because some insecticides are toxic to fish, you must not spray lakes, streams or ponds. Do not pollute water by cleaning equipment or disposing of waste insecticide in any body of water.

You can spray for ticks any time of the year, but for the treatment to be most effective, you should treat when ticks are most likely to be active in your area. A late-spring treatment will kill ticks that have survived the winter. A late-summer treatment will kill larvae that have hatched from eggs, and nymphs and adults that have molted from larvae and nymphs that fed during the summer. You may have to treat at intervals throughout the summer if tick numbers are high and there is a threat of disease in your area. A fall treatment will decrease the number of ticks that overwinter and will be looking for hosts the next spring.

If you suspect you are moving into a shelter or a building that may harbor relapsing fever ticks, you may wish to use one of the insecticides approved for the control of ticks in houses.

A final important note: If you use insecticides to control ticks, you must follow the label instructions exactly. The label is there to provide you with information so you can use these products safely and effectively.

Chapter 10

Other Tick Control Methods

In addition to using insecticides, you may be able to use other methods to control ticks. You may find you can use some of these methods, while others are expensive and not practical on a small scale. However, this book describes these other methods so you will know about them in case you are able to use them.

Insecticide-Treated Nest Material

A product called Damminix is used to control the deer tick. It consists of a tube that contains cotton treated with permethrin—the same insecticide found in Permanone™ Tick Repellent. Damminix works by killing larvae and nymphs on mice. Mice live on the ground and make nests out of leaves, grass, etc. Mice carry the cotton to their nests, where the insecticide kills ticks on the mice.

Damminix works best when you place tubes where mice live. The label calls for placing tubes at 10-yard intervals. This spacing is critical, for it puts tubes within the home range of the mice. Place tubes on the ground at the edges of lawns, in gardens, and in brushy areas near open spaces. The cotton should stay dry, for mice prefer to nest in dry, fluffy material.

You should treat with Damminix twice each year. The first application, in April or May, kills the overwintered nymphs; the second, in July, kills the larvae that have hatched in the summer. The recommended treatment of 48 tubes per acre reduced the number of infested mice and the number of ticks on mice by 97%. The number of nymphs the next spring was reduced by 97%.

Damminix is approved by the EPA and in 1989 was registered in 21 states—Connecticut, Delaware, Georgia, Illinois, Indiana, Kentucky, Maryland, Massachusetts, Michigan, Minnesota, New Hampshire, New Jersey, New York, North Carolina, Ohio, Pennsylvania, Rhode Island, South Carolina, Vermont, Virginia and Wisconsin. Registration was pending in California, Missouri and Maine.

A somewhat similar treatment has been tested against the American dog tick. This technique, called the baited pesticide treatment station, consists of a short piece of PVC pipe that contains felt strips treated with an insecticide. Peanut butter, placed inside the pipe, is a food bait to attract mice. Mice enter the pipe and contact the insecticide. When treated pipes were placed in the field in the spring to control overwintered larvae and nymphs, the numbers of American dog ticks on mice were reduced by 81-98%.

Vegetative Management

You can manage vegetation to control ticks, especially the lone star tick. The purpose of vegetative management is twofold. The first is to decrease areas covered with shrubs and undergrowth, which are the favorite habitat of tick hosts. The other is to remove the leaf litter and duff on the soil, to eliminate the protective layer where ticks live. Sunlight reaching the ground dries the soil, making it unsuitable for eggs and other stages of the lone star tick.

To decrease the numbers of the deer tick, you can reduce the underbrush and remove the trash from your yard to lessen its attractiveness as nesting sites for the mice that are hosts of the deer tick.

You can manage vegetation mechanically by trimming or removing trees to thin out the overstory, removing the understory vegetation (shrubs, tall weeds, etc.) and closely mowing the grass. You can reduce or remove leaf litter and duff by controlled burning. Fire will consume the leaf litter and duff and also kill ticks in that layer and in the soil. You can also manage vegetation by applying EPA-approved herbicides according to label instructions.

These different methods of vegetation management require different amounts of money, knowledge and experience. The easiest to use in a small area is mechanical clearing: trim or cut down certain trees by hand, eliminate understory vegetation with a weed cutter and mow grasses closely. An effective controlled fire takes a great deal of knowledge, skill and preparation. You must have enough dead grass, leaves and other burnable material on the soil to support an effective fire. A fire that burns too quickly and without enough heat will not kill ticks. Fires get out of control easily and can be very dangerous. Herbicides can be effective, but you must use them with caution: you need knowledge, experience and proper equipment to use herbicides correctly and safely.

Host Management

Since ticks that carry diseases feed on wild animals, especially adult ticks on deer, to get rid of or decrease the number of ticks, all you need to do is get rid of or decrease the number of deer.

Researchers removed about 70% of the deer from an island in Nantucket Sound,

Mass. and saw no reduction in the number of larvae of the deer tick on mice the next summer. They removed all the deer and saw a rapid decrease in the number of larvae. Because the life cycle of the deer tick takes 2 years, it takes a while to see the effects of the removal of deer on the numbers of this tick.

In studies with the lone star tick, researchers built a deer-proof fence around an area near Golden Pond, Kentucky. Afterward, they saw a 98% decrease in the number of larvae inside the area but only a small decrease in the number of adults and nymphs. They said this lack of reduction was due to the fact that these stages could have fed on medium-sized wild animals—raccoons, foxes, skunks, etc., not excluded by the fence. They found an increase in the number of American dog ticks. In other tests, they built a deer-proof fence around a large campsite and also used vegetative management and insecticides in limited areas of the campsite. Afterwards, they saw a reduction in the number of larvae and nymphs of the lone star tick inside the deer-free, treated area, but no decrease in the number of American dog ticks.

Deer are accepted and wanted wildlife, and it seems unlikely people will give up deer. An alternative is deer-proof fences. They are expensive to construct and maintain, but the cost may be justified if ticks or the diseases they carry are reduced or eliminated.

A word should be said about integrating insecticide application (Chapter 9) with the techniques listed above. Each technique has its plusses and minuses. Insecticide application gives fast control but is expensive and must be done correctly to be effective and safe. You may need to repeat insecticide treatments yearly or more often. Vegetation management is less effective and slower acting, but is longer lasting. You may need to repeat vegetation management every third or fourth year. Host management is the slowest-acting, and it may be necessary to eliminate all the deer to affect the number of ticks.

In any area, it may be possible to combine one or more of these techniques to decrease the number of ticks. For example, in one area you may wish to apply insecticides, while in another it may be possible to combine the use of Damminix with vegetation management to reduce numbers of ticks. A great deal depends upon the magnitude of the tick problem, the ticks present, whether disease is a threat and how much money, time and knowledge are available for tick control. Integrating these techniques has the potential to reduce the number of ticks and thus the diseases and other problems caused by ticks.

List of Illustrations

GOPRO INVENTOR NICK WOODMAN

MATT DOEDEN

Lerner Publications
Minneapolis

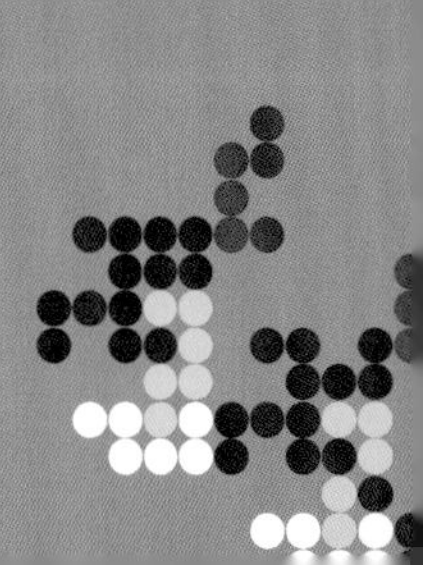

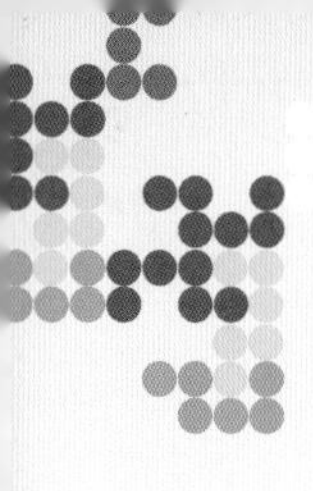

Lerner Publications Company
A division of Lerner Publishing Group, Inc.
241 First Avenue North
Minneapolis, MN 55401 USA

For reading levels and more information, look up this title at www.lernerbooks.com.

Content Consultant: Tim Meehan, instructor and consultant in digital arts, and certified paragliding instructor for FlyTim

The Cataloging-in-Publication Data for *GoPro Inventor Nick Woodman* is on file at the Library of Congress.
ISBN 978-1-4677-5792-8 (lib. bdg. : alk. paper)
ISBN 978-1-4677-6117-8 (pbk.)
ISBN 978-1-4677-6281-6 (EB pdf)

Manufactured in the United States of America
1 – PC – 12/31/14

The images in this book are used with the permission of: © iStockphoto.com/EpicStockMedia, p. 4; K.C. Alfred/ZUMA Press/Newscom, pp. 5, 15, 26; © Barry Winiker/Photolibrary/Getty Images, p. 6; © Jim Feliciano/Shutterstock.com, p. 7; © Bloomberg via Getty Images, pp. 9, 16, 18, 25; © iStockphoto.com/Henrik5000, p. 10; © iStockphoto.com/MyfanwyJaneWebb, p. 12; © LatitudeStock/Alamy, p. 13; © Svetlana Sysoeva-Fotolia.com, p. 14; AP Photo/PR NEWSWIRE, p. 19; © EMMANUEL FOUDROT/Reuters/CORBIS, p. 20; © Skip Brown/National Geographic/Getty Images, p. 21; © Denis Poroy/Getty Images, p. 22; © MIKE SEGAR/Reuters/CORBIS, p. 24; Red Bull Stratos/EPA/Newscom, p. 27.

Front Cover: © Bloomberg via Getty Images (camera); Max Morse/Wikimedia Commons (CC BY 2.0).

Main body text set in Adrianna Regular 13/22. Typeface provided by Chank.

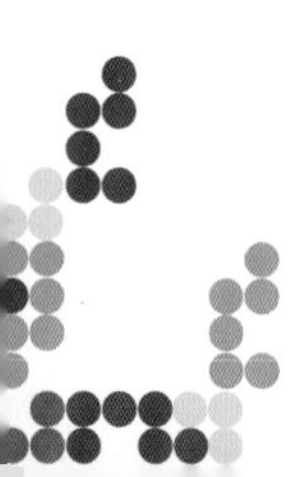

CONTENTS

Chapter 1

Nick Woodman grew up near the California coast, where surfing is popular. Like this surfer, Woodman enjoyed catching waves.

SURFER BOY

Plenty of technology lovers dream of living in California's Silicon Valley. Nick Woodman didn't have to. He grew up there. Born in 1975, he witnessed the rise of giant tech companies such as Apple and Microsoft. Yet as a boy, Woodman never dreamed of taking the tech world by storm.

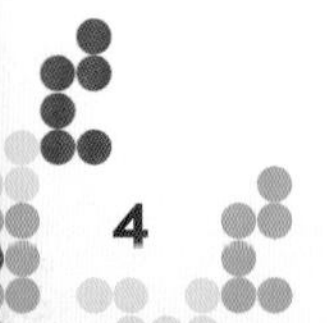

Woodman's goal was far simpler. He knew what he wanted to do from the time he was eight years old. He wanted to surf.

The youngest of four children, Woodman was a natural athlete. He played baseball and football in high school. But by the time he was a senior, surfing had become his greatest love. In his free time, he often drove 45 miles (72 kilometers) to the nearest beach. He even started a surfing club at his high school.

Woodman's passion for surfing helped change the world of cameras.

CHASING WAVES

Woodman was a good student. And he had a knack for technology. When he wasn't surfing, he was building remote-controlled gliders. But when he graduated from high school in 1993, Woodman wasn't planning on a career in technology. He was thinking mostly about his next wave. That's why he wanted to attend the University of California, San Diego (UCSD). It was close to one of the world's best surfing beaches, Black's Beach.

Surfers know Black's Beach for its large and powerful waves.

Woodman attended the University of California, San Diego.

But UCSD denied Woodman's application. Instead, the University of California, Berkeley, accepted him. Woodman's parents begged him to go there. Berkeley was too far from the beach, though. Woodman was determined to go to UCSD. So he wrote an appeal to the school, explaining why he was a good fit. The university's officials changed their minds. Woodman was in!

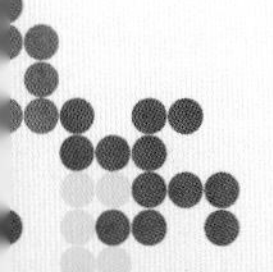

San Diego was a perfect spot for him and his surfing classmates. Between classes, he spent as much time as he could at the beach. He hadn't come to UCSD just to ride waves, though. He'd also come to learn.

Woodman started out studying economics. But he didn't enjoy the classes. So he followed his passions. He focused on subjects such as art, writing, and acting. He went on to earn a degree in visual arts.

Woodman also found love while at UCSD. He met a fellow student, Jill Scully. They hit it off, began dating, and eventually married.

TECH TALK

"In a world where we all try to figure out our place and our purpose here, your passions are one of your most obvious guides. They lead you along life's path . . . to your career."

—Nick Woodman

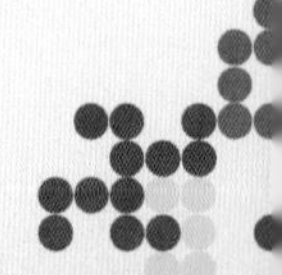

Woodman pursued a degree in visual arts. He continues to use his creative talents to develop new products.

Chapter 2

Many new Internet companies were formed in the late 1990s.

FINDING A PLACE

Woodman had big plans after college. He tried to start a website that would sell electronics. His plan was to price every item at two dollars or less. But the business venture failed almost immediately.

Woodman quickly moved on to his next idea. In 1999, he started a company called FunBug. FunBug was a website that combined video games and contests. Users could play games

to win cash prizes. This time, Woodman managed to sell his idea to several big **investors**. He and his investors poured almost $4 million into FunBug.

At the time, the Internet was exploding into a big business. Plenty of young tech wizards were striking it rich. All a person seemed to need was a winning idea. Woodman hoped that FunBug was just such an idea.

Unfortunately for Woodman, it wasn't. Part of the problem was timing. Internet companies had thrived in the late 1990s, but many of them collapsed in 2000 and 2001. The crash marked the end of the "dot-com boom." FunBug was one of its many victims.

DOT-COM BOOM . . . AND BUST

The dot-com boom was a period of rapid growth for Internet-related businesses. In the late 1990s, hundreds of tech companies sprang up. By 2000, many of these companies began to fail. Some companies had poor business plans. Some couldn't compete with similar companies. Only a small fraction of the original dot-com businesses survived past the early 2000s.

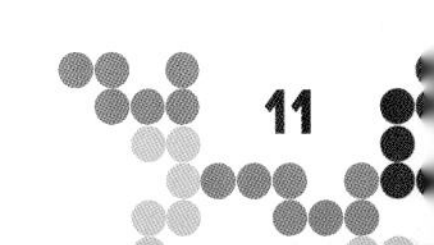

A surfer rides a wave in Australia. Woodman was inspired to start a new business during a surfing trip to Australia and Indonesia.

BACK TO THE BEACH

Twice Woodman had tried to launch his own business. Twice he had failed. He began to question whether he had the skill to succeed.

"I'd never failed at anything before except computer science engineering classes," he said. "So it was like . . . maybe I'm not capable of doing this."

Woodman needed to clear his head. It was time to surf. In 2002, he took a trip to Australia and Indonesia. There, he would ride some of the world's best waves.

Like many surfers, Woodman wanted to capture some of his greatest waves on camera. But he wasn't eager to surf while holding a camera. He wanted the use of both of his hands to stay balanced. And he didn't want to lose his camera at the bottom of the ocean.

Woodman had an idea. He attached a disposable camera to an old wrist strap from a surfboard. Then he wrapped the strap around his wrist. If the strap stayed in place, he'd be able to carry the camera and still keep his hands free. To snap a photo, he'd only need to hold up his arm and press the camera's top button. An instant later, both hands would be free again.

Woodman hit the beach to test out his invention. He paddled out, found his perfect wave, and snapped a picture. It worked! For five months, Woodman put his invention to the test in the waters of the South Pacific. And he started to form a plan.

Disposable cameras are easy to replace if they are broken or lost. These types of cameras were popular in the late 1990s.

TRY, TRY AGAIN

Woodman returned home energized by his trip. The surfing had helped him refocus on his future. He was ready to improve his invention. The camera strap worked, but it wasn't very secure. The camera tended to flop around loosely on his wrist.

Woodman wanted to upgrade his strap and sell it to other surfers. But first, he needed money. The materials to make the strap were expensive. And Woodman didn't have much savings left.

But Woodman had figured out how to raise some cash. While on his trip, he had bought hundreds of bead-and-seashell belts from a market in Indonesia. He'd bargained the seller

To raise money for his new invention, Woodman and his wife, Jill, sold bead-and-seashell belts like these.

Woodman takes a photo with his wife, Jill, during an alumni event at UCSD. The couple met when they were students at the college.

down to less than two dollars per belt. Then he brought them home. He and Jill loaded up their car and drove up and down the California coast, selling the belts. They sometimes got as much as sixty dollars per belt.

Woodman's family helped out by giving him a **loan**. Finally, Woodman had enough money to start his new business. He called it Woodman Labs. But soon, it would be known by the name of its products: GoPro.

Chapter 3

Woodman shows off the Hero3+ model. This version was released in 2013.

MAKING A SPLASH

Woodman knew that straps were just the beginning. He needed to create the full package. That included a strap, a camera, and a sturdy casing. At first, Woodman planned to use existing camera models. But he needed

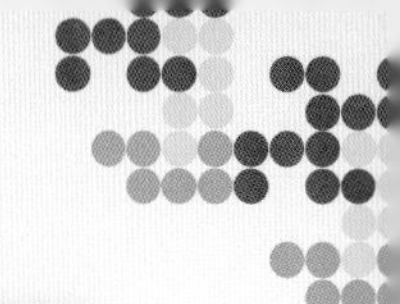

TECH TALK

"Every camera I used would flood or break after a big wipeout. I realized I shouldn't be a strap company but a camera company."

—Nick Woodman

something that would hold up in tough conditions. Big-name cameras were too fragile. "I went to all the major camera shows, walking every aisle," he said. "I looked at every booth twice." Yet nothing was a good fit.

Woodman needed a plan B. He found a little-known Chinese manufacturer called Hotax. This company made reusable cameras for snorkelers. Hotax agreed to make cameras for Woodman. And then he was off and running.

Woodman locked himself inside his home and got to work. He often spent eighteen hours a day working with drills, his mother's sewing machine, and other tools. Woodman left nothing to chance. He wore his own strap everywhere, even to bed. He had to know if the strap had any weak points or would make someone sore. And the only way to know was to use it—constantly.

THE HERO

In 2004, Woodman's system was finally ready to go. He went to an action sports convention in San Diego with his first finished model, the GoPro Hero. This early model included a strap with a 35mm **film** camera. It was a hit. One Japanese buyer ordered one hundred of the gadgets, paying more than $2,000.

From there, Woodman's ideas only got bigger. He wanted to make sure that GoPro was cutting edge. Soon a **digital** camera replaced the film version. Next came a camera that could record short bursts of video.

Woodman holds an original waterproof GoPro camera that used film.

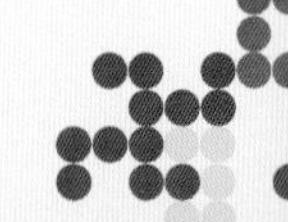

This woman uses a digital GoPro Hero while diving.

GoPro was growing and growing. Woodman hired several college friends—and Jill—as his first employees. But by 2007, Woodman feared he had taken the company as far as he could. He planned to sell control of GoPro to an investment group.

Then in 2008, the world suffered a terrible economic **recession**. The investors who wanted to buy the company lowered their offering price. Woodman refused to sell.

The Hero HD can be mounted on a variety of helmets.

HI-DEF

GoPro wasn't just for surfers anymore. It had quickly become big in the larger action sports community. In 2009, the company really got athletes' attention. The new Hero HD included a **high-definition** video camera. Anyone—from a race car driver to a skydiver—could record a crystal-clear video of a memorable moment. The Hero HD cost less than most high-quality camcorders too.

The Hero HD was unlike any other camera. It was a smash hit. And that wasn't because of a huge advertising campaign. Much of the success came from word of mouth, especially by social media. Soon athletes were posting their videos to Internet sites such as YouTube.

This photo was captured by a GoPro camera mounted to the front of the paddleboard.

At a 2014 Major League Baseball game, Woodman gets ready to throw out the first pitch. A GoPro camera attached to his cap records the event.

Many of these GoPro videos drew thousands or even millions of views. Athletes tried to post better, more daring videos. Others tried to copy them. And sales soared. Within a year, the Hero HD brought in more than $60 million. Woodman's company had hit the big time.

For Woodman, the experience was amazing. He hadn't expected this level of success. He'd simply worked hard and had fun, as he did with surfing. And as GoPro grew, Woodman kept looking for ways to improve his products. He sensed that the company's journey was only beginning.

Chapter 4

Starting in 2014, members of the general public could invest in GoPro. Public investments can increase a company's value. Woodman and his team celebrated the day GoPro began taking on public investors.

LOOKING FORWARD

Woodman had started his company as a way for surfers to take close-up action photos. But GoPro had grown into much more. The company began making mounts that

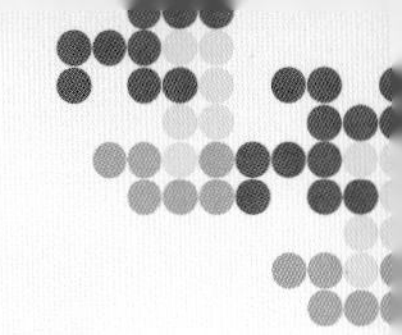

allowed users to put the camera anywhere. Soon the Hero HD had been from outer space to the bottom of the ocean. It was on Hollywood sets and on-site for TV reality shows. One even went deep down into a collapsed mine shaft in Chile!

THE MAD BILLIONAIRE

The 2008 recession had little effect on Woodman's company. GoPro continued to grow. In 2010, the electronics chain Best Buy began to carry the devices in its stores. Once again, sales soared.

GoPro has developed a wide selection of mounts and straps for cameras.

Woodman *(center)* talks with students after giving a speech in 2014.

In 2012, Woodman got an offer he couldn't refuse. An investment group bought less than 9 percent of his company for a stunning $200 million. Based on that price, GoPro was worth more than $2 billion! A few months later, *Forbes* magazine listed thirty-seven-year-old Woodman as one of the youngest billionaires in the world.

Yet Woodman's great wealth didn't change his passions. He still loved to surf, snowboard, ski, bungee jump, and do other action sports. Some members of the media nicknamed him the Mad Billionaire, due to his extreme hobbies!

Woodman and GoPro continued to draw attention. In 2012, daredevil Felix Baumgartner used one of the company's cameras when he set a record for the highest skydiving jump. A 2014 Super Bowl commercial aired some of the footage. GoPro was rapidly becoming a household name.

BIG PLANS

Woodman plays hard. But he also works hard. And he's not yet done building GoPro. Recently, he's begun designing cameras that can stream video online in real time. Viewers can watch an event as it happens!

Felix Baumgartner wore a GoPro camera during his record-setting skydive in 2012.

Action sports remain a focus for GoPro. But Woodman and his team are always looking for new customers. Surgeons, law enforcement officers, and scientists are among GoPro's users.

In less than ten years, GoPro went from nothing to the fastest-growing camera company in the United States. Who knows where it could be in another ten years? No matter where it goes, Woodman will likely enjoy the ride.

TECH TALK

"In surfing terms, there are some waves . . . where you start off easy. But as you go down the line, the wave keeps growing, and the barrel keeps getting bigger and bigger. You keep going faster and faster and faster. When people ask me what it's like [running GoPro], I say it's kind of like that. For my friends and me, this is the best ride of our lives."

—Nick Woodman

TIMELINE

1975
Nicholas Woodman is born in California.

1993
Woodman begins attending college at the University of California, San Diego.

1999
Woodman starts an Internet game site called FunBug, which soon fails.

2002
Woodman founds Woodman Labs, later known as GoPro.

2004
Woodman unveils the GoPro Hero, which includes a 35mm still camera.

2009
GoPro launches the Hero HD. The brand's popularity explodes.

2010
Consumer electronics giant Best Buy begins carrying GoPro models.

2012
Woodman sells less than 9 percent of GoPro for $200 million. Felix Baumgartner uses a GoPro camera to film his world record for the highest skydiving jump.

2013
Forbes declares Woodman one of the world's youngest billionaires.

2014
Woodman Labs officially changes its name to GoPro. Felix Baumgartner's GoPro footage from his record-breaking skydiving jump appears in a Super Bowl commercial.

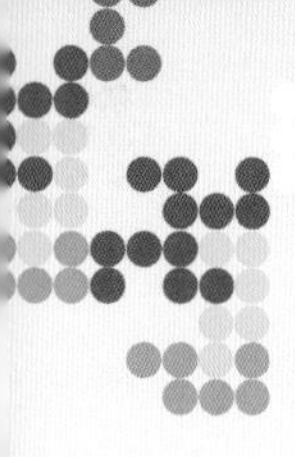

SOURCE NOTES

8 Ryan Mac, "Five Startup Lessons from GoPro Founder and Billionaire Nick Woodman," *Forbes*, March 13, 2013, http://www.forbes.com/sites/ryanmac/2013/03/13/five-startup-lessons-from-gopro-founder-and-billionaire-nick-woodman.

12 Ryan Mac, "The Mad Billionaire Behind GoPro: The World's Hottest Camera Company," *Forbes*, March 4, 2013, http://www.forbes.com/sites/ryanmac/2013/03/04/the-mad-billionaire-behind-gopro-the-worlds-hottest-camera-company.

17 Robert Moritz, "Guts, Glory, and Megapixels: The Story of GoPro," *Popular Mechanics*, June 12, 2012, http://www.popularmechanics.com/outdoors/sports/technology/guts-glory-and-megapixels-the-story-of-gopro-8347639.

17 Ibid.

28 Serena Renner, "The Best Ride of Our Lives," *Triton*, accessed May 14, 2014, http://alumni.ucsd.edu/s/1170/emag/emag-interior-2-col.aspx?sid=1170&gid=1&pgid=4381.

GLOSSARY

digital
using or characterized by computer technology

film
a light-sensitive medium used to capture images in cameras

high-definition
a digital video standard that features a very high degree of detail

investors
people or companies that put money into a company or a project in return for a share of the profits later

loan
money that is given to someone with the agreement that it will be paid back later

recession
a widespread economic downturn

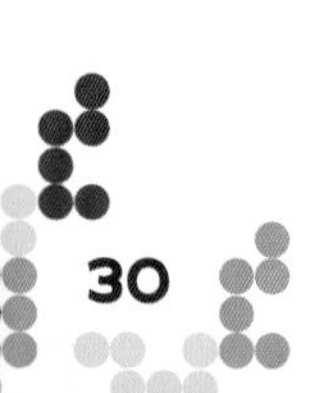

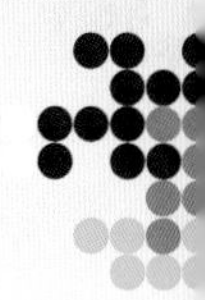

FURTHER INFORMATION

BOOKS

Doeden, Matt. *SpaceX and Tesla Motors Engineer Elon Musk.* Minneapolis: Lerner Publications, 2015. Learn about another young tech lover whose business ventures made him a billionaire.

Hubbard, Ben. *Hi-Tech World: Cool Stuff.* New York: Crabtree, 2010. Explore more cutting-edge tech inventions.

Kallen, Stuart A. *Digital Cameras and Camcorders.* Detroit: Lucent Books, 2014. Discover the technology behind cameras such as GoPro.

WEBSITES

GoPro
http://gopro.com
The official website for Nick Woodman's camera company shows its latest products.

How High-Definition Camcorders Work
http://electronics.howstuffworks.com/cameras-photography/digital/high-definition-camcorder.htm
Find out more about how high-tech cameras work.

YouTube—GoPro
https://www.youtube.com/user/GoProCamera
Check out videos of amazing stunts and faraway places taken with GoPro cameras.

Expand learning beyond the printed book. Download free, complementary educational resources for this book from our website, www.lerneresource.com.

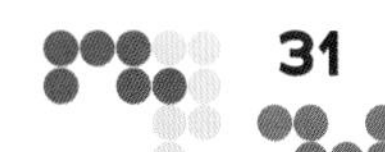

INDEX

ABOUT THE AUTHOR

Matt Doeden studied journalism at Mankato State University, where he worked at the college newspaper for three years. Doeden went on to work as a sportswriter for the Mankato paper and then got a job as an editor for a children's book publisher. In 2003, Doeden decided to start his own business as a freelance writer and editor. Since then, he has written and edited hundreds of books on high-interest topics such as cars, sports, and airplanes, as well as curricular topics such as geography, science, and math.